I0068682

Published by New Generation Publishing in 2013

Copyright © Darren St. John Sumner 2013

First Edition

The author asserts the moral right under the Copyright, Designs and Patents Act 1988 to be identified as the author of this work.

All Rights reserved. No part of this publication may be reproduced, stored in a retrieval system or transmitted, in any form or by any means without the prior consent of the author, nor be otherwise circulated in any form of binding or cover other than that which it is published and without a similar condition being imposed on the subsequent purchaser.

www.newgeneration-publishing.com

New Generation Publishing

G-500/G-501:-
The First Force of Movement Is

MOVEMENT

Movement is one body moving around
other bodies.
Therefore one body is separate
from other bodies to do this.

G-500/G-502:-
The Second Force of Movement Is

MOVEMENT

MOMENTUM

MOMENTUM

MOMENTUM

MOMENTUM

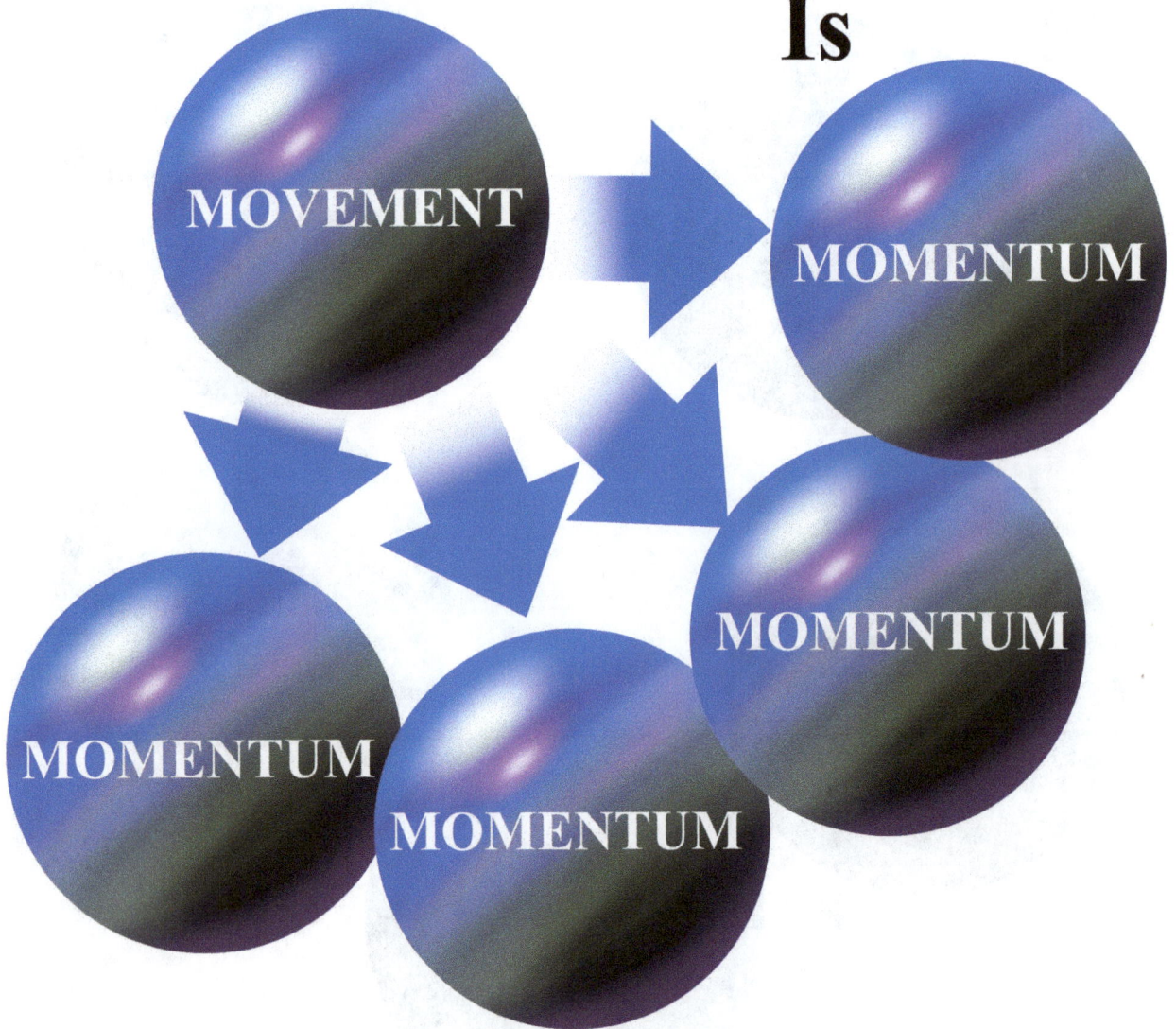

Momentum is other bodies being
pushed forwards by one body
to be assembled into shapes and patterns.
Therefore one body interacts
with other bodies to do this.

G-500/G-503:-
The Third Force of Movement Is

MOVEMENT

MOBILE

MOBILE

MOBILE

MOBILE

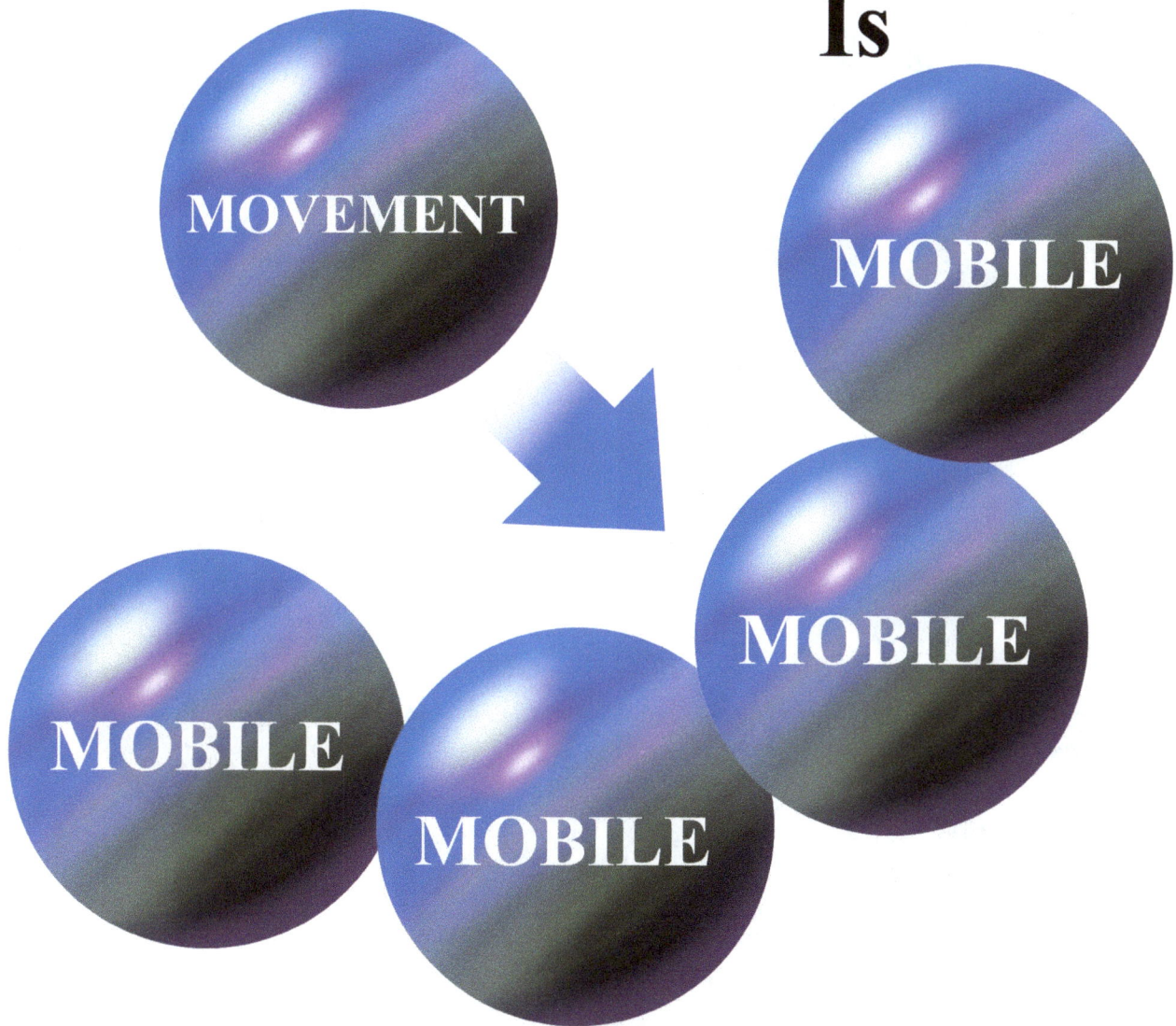

Mobile is assembled shapes and patterns being pushed forwards by one body. Therefore assembled shapes and patterns move away from one body.

G-500/G-504:-
The Fourth Force of Movement Is

MOVEMENT

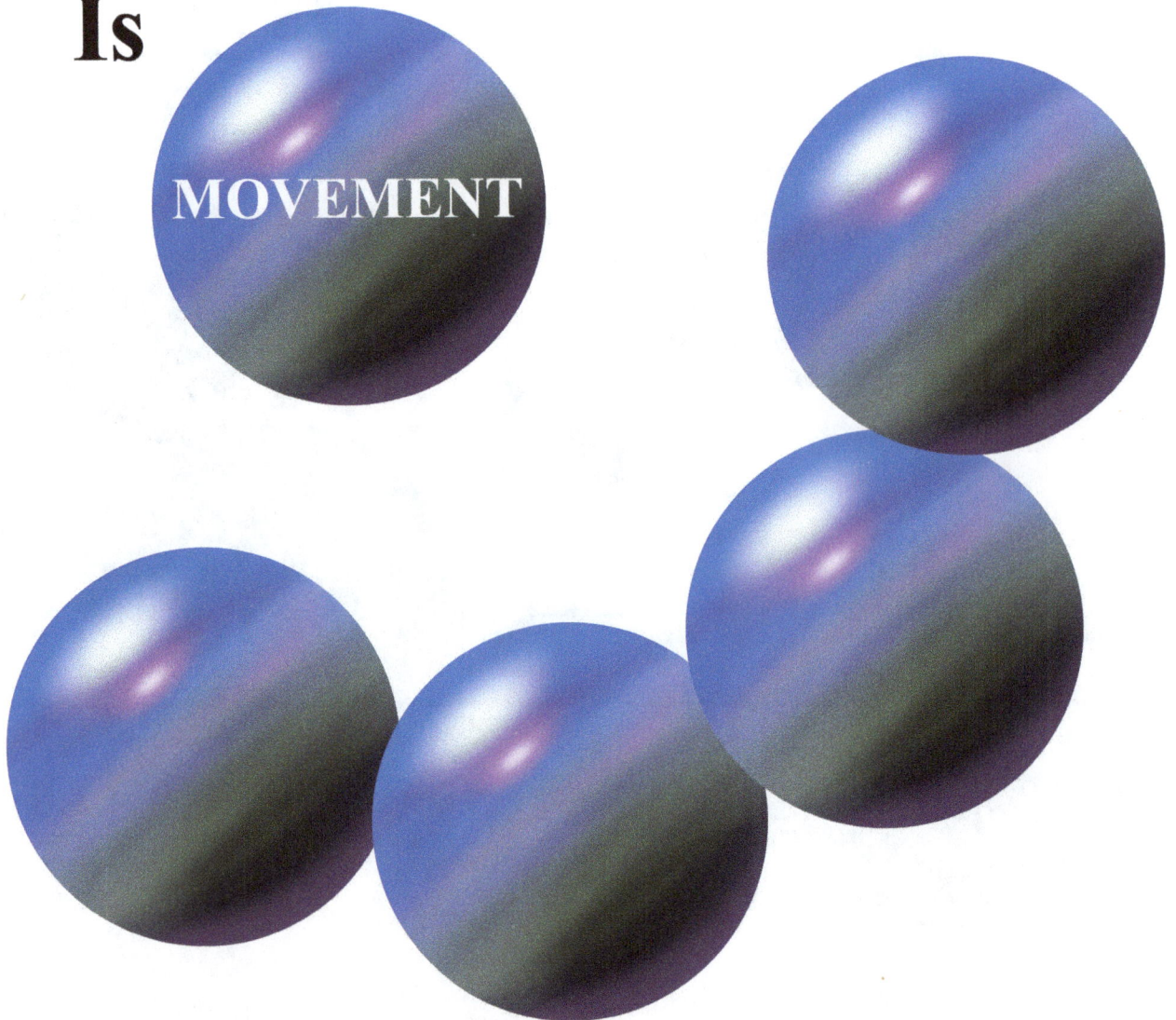

Movement is one body moving around
assembled shapes and patterns.
Therefore one body is separate from
assembled shapes and patterns to do this.

G-500/G-505:-
The Fifth Force of Movement Is

MOVEMENT

MOMENTUM

MOMENTUM

MOMENTUM

MOMENTUM

MOMENTUM

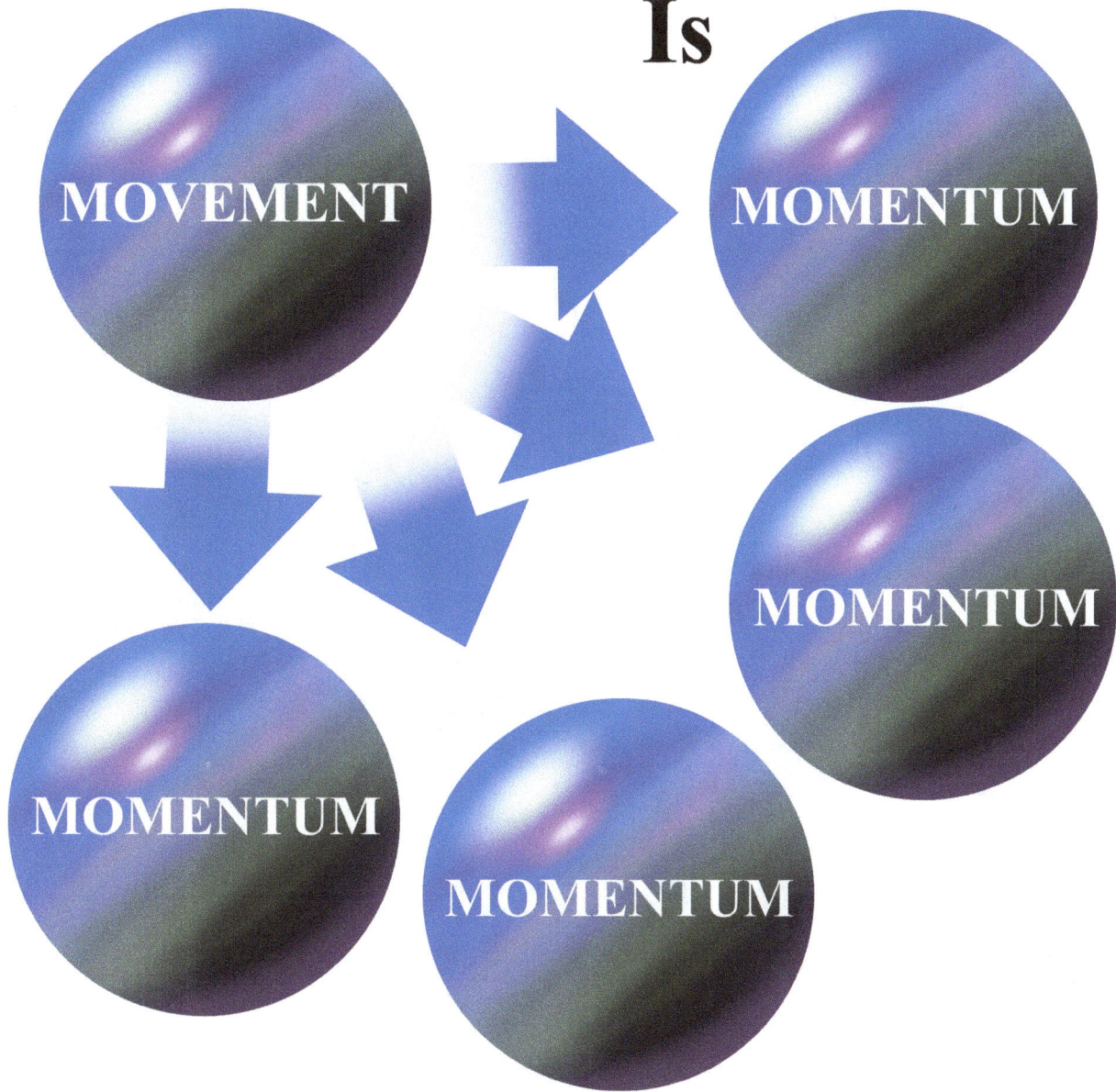

Momentum is assembled shapes and patterns being pushed apart forwards by one body. Therefore one body interacts with assembled shapes and patterns to do this.

G-500/G-506:-
The Sixth Force of Movement Is

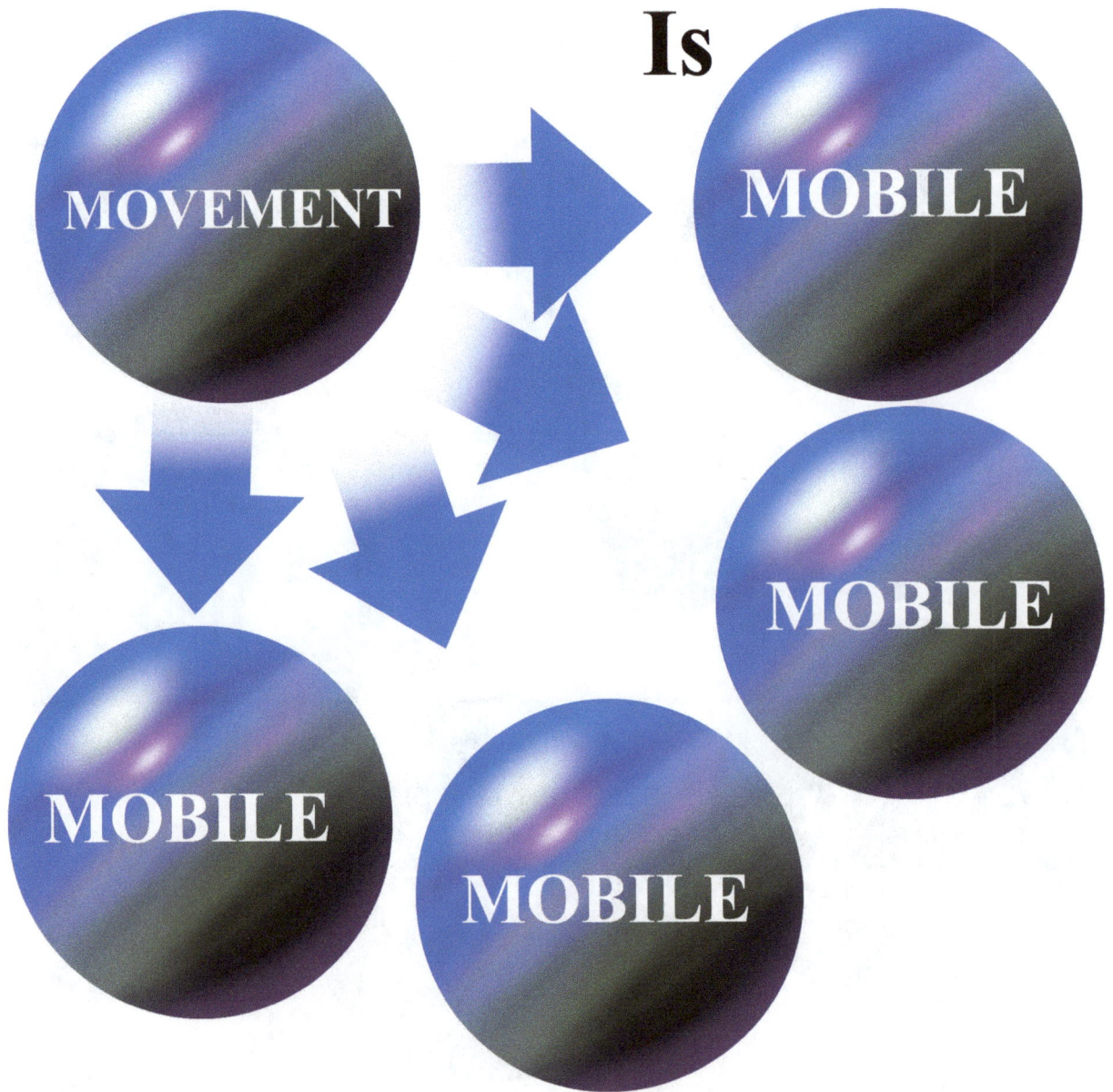

MOVEMENT

MOBILE

MOBILE

MOBILE

MOBILE

Mobile is parted shapes and patterns being pushed forwards by one body. Therefore parted shapes and patterns move away from one body.

G-500/G-507:-
The Seventh Force of Movement Is

MOVEMENT

Movement is one body moving around
other bodies.
Therefore one body is separate
from other bodies to do this.

G-500/G-508:-
The Eighth Force of Movement
Is

MOVEMENT

MOTION

MOTION

MOTION

MOTION

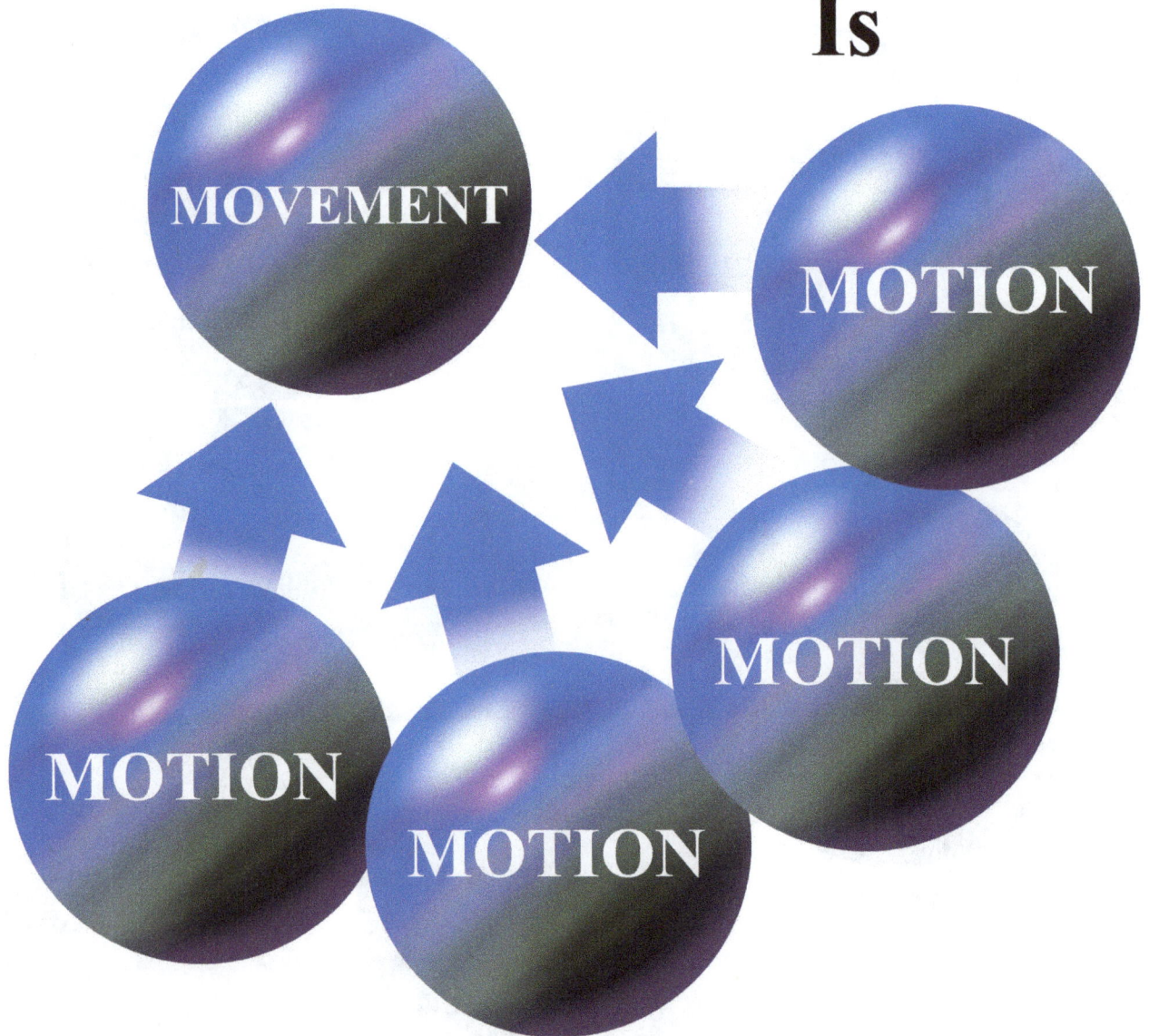

Motion is other bodies being pulled backwards
to be assembled into shapes and patterns
by one body.
Therefore one body interacts
with other bodies to do this.

G-500/G-509:-
The Ninth Force of Movement Is

MOVEMENT

MOBILE

MOBILE

MOBILE

MOBILE

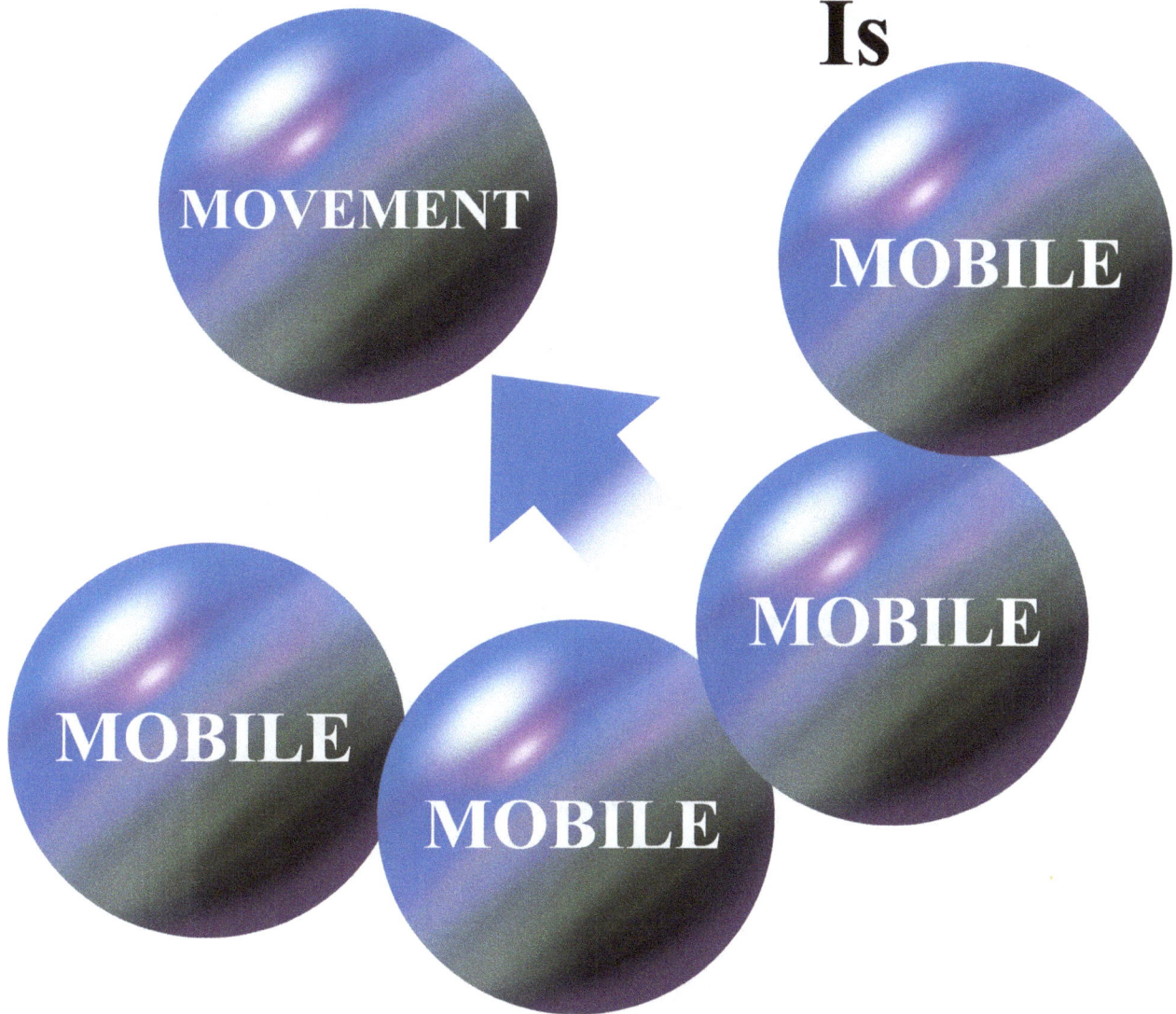

Mobile is assembled shapes and patterns being pulled backwards by one body. Therefore assembled shapes and patterns join with one body.

G-500/G-510:-
The Tenth Force of Movement Is

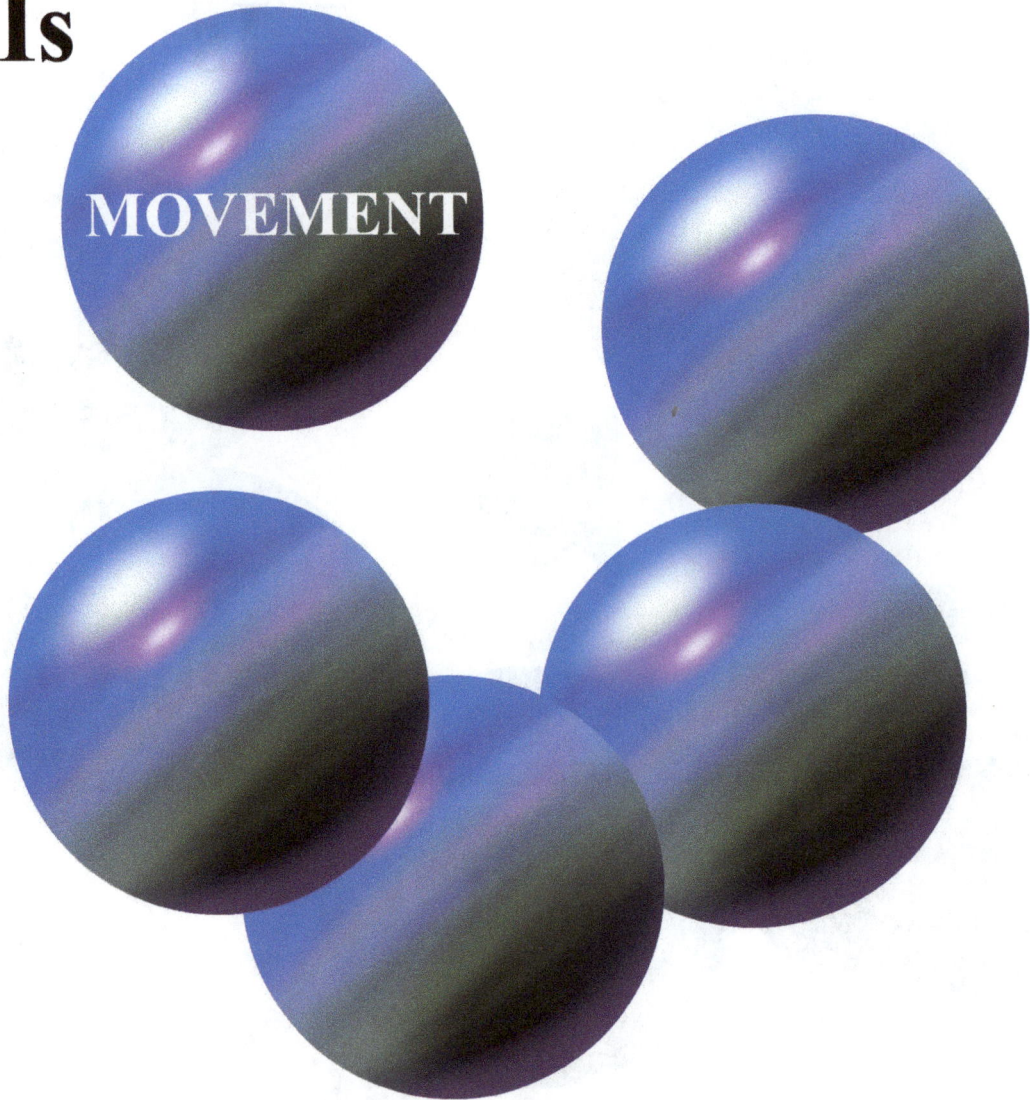

MOVEMENT

Movement is one body moving around assembled shapes and patterns. Therefore one body is separate from assembled shapes and patterns to do this.

G-500/G-511:-
The Eleventh Force of Movement Is

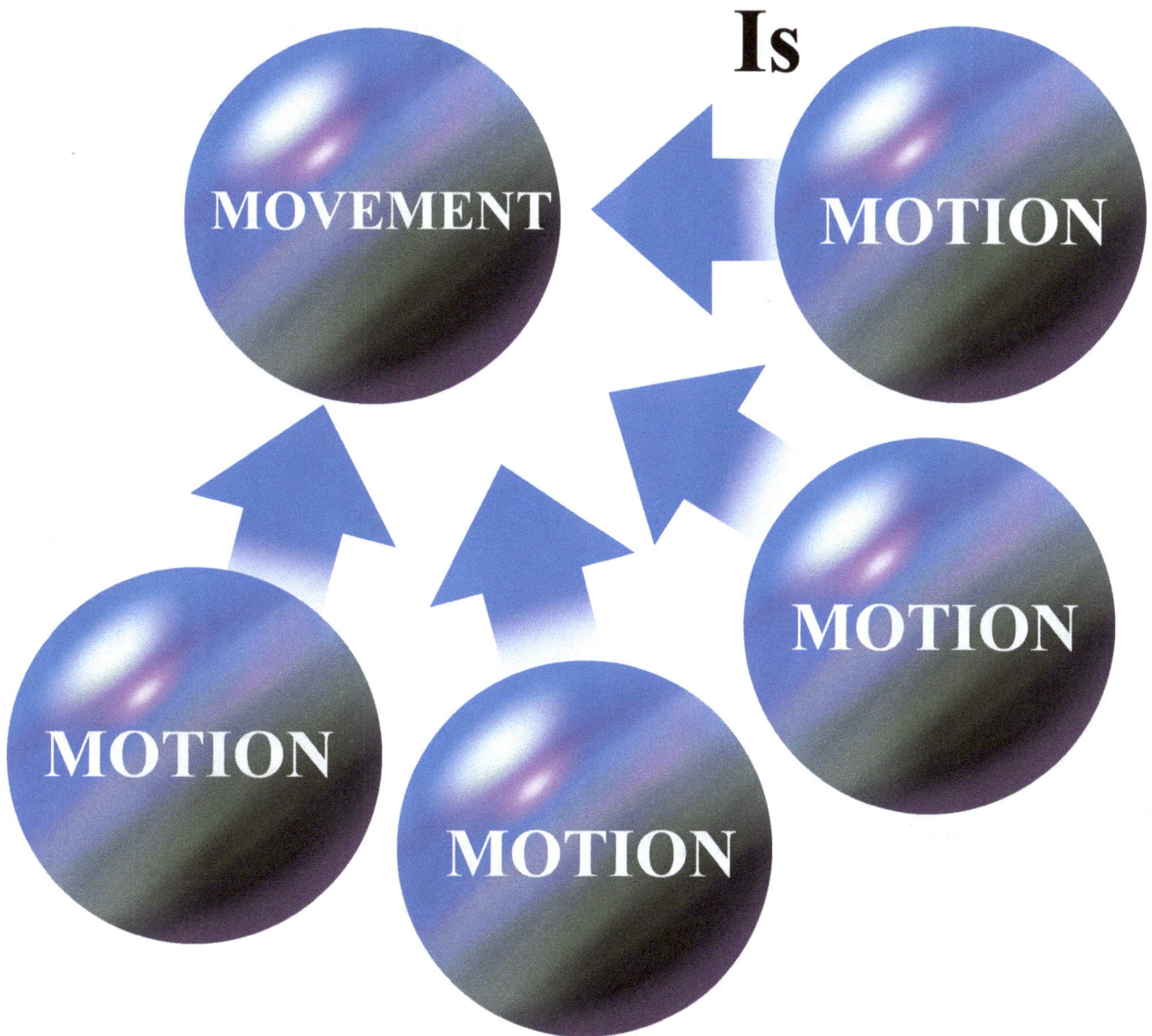

Motion is assembled shapes and patterns being pulled apart backwards by one body. Therefore one body interacts with assembled shapes and patterns to do this.

G-500/G-512:-
The Twelfth Force of Movement Is

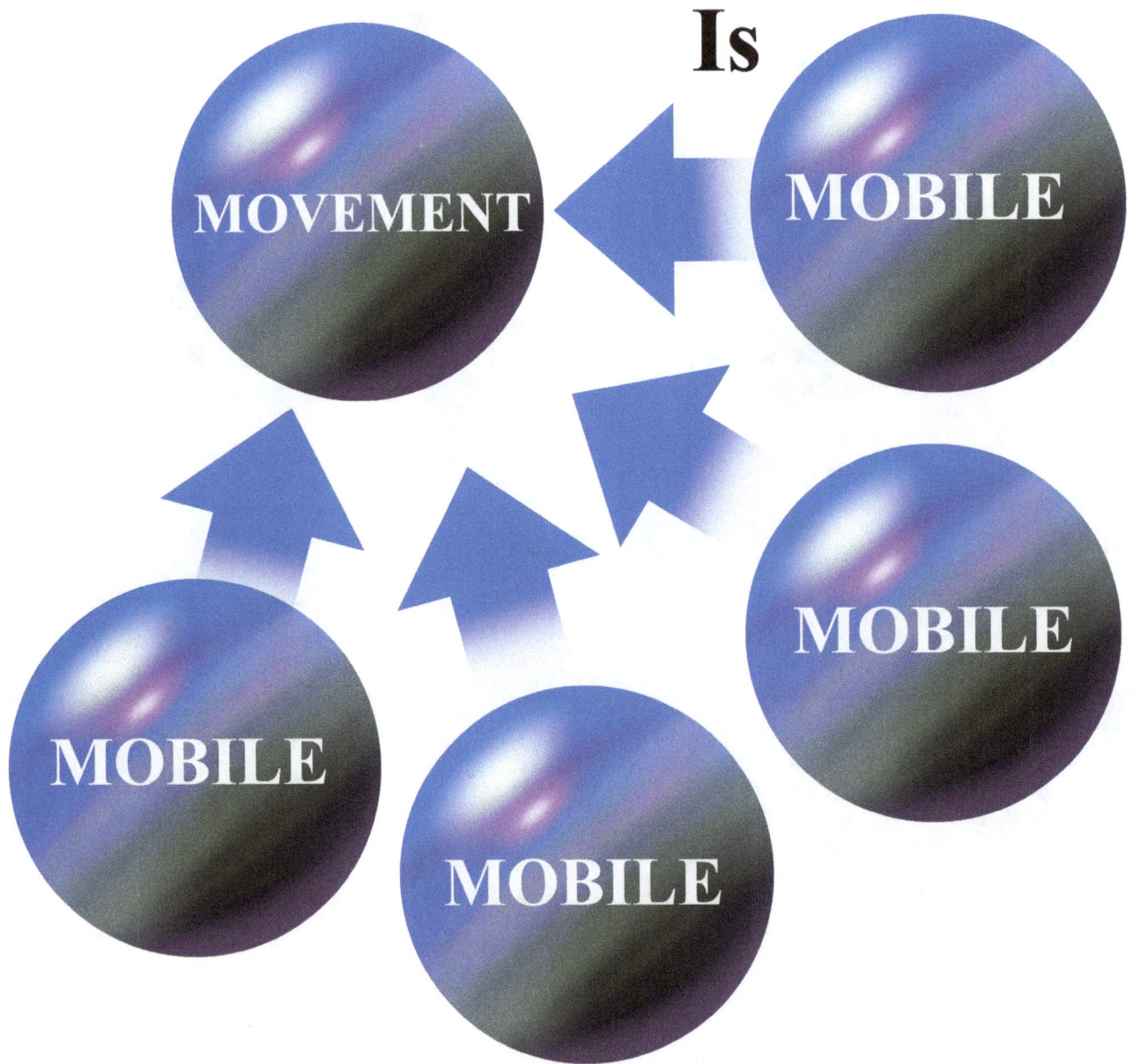

MOVEMENT

MOBILE

MOBILE

MOBILE

MOBILE

Mobile is parted shapes and patterns being pulled backwards by one body. Therefore parted shapes and patterns join with one body.

www.ingramcontent.com/pod-product-compliance
Lightning Source LLC
Chambersburg PA
CBHW082303210326
41521CB00035B/2456